Frederic Russell Sturgis

Human Cestoids

Frederic Russell Sturgis

Human Cestoids

ISBN/EAN: 9783337365585

Printed in Europe, USA, Canada, Australia, Japan

Cover: Foto ©berggeist007 / pixelio.de

More available books at **www.hansebooks.com**

AN ESSAY

ON

THE HUMAN CESTOIDS.

Human Cestoids:

AN ESSAY

TO WHICH WAS AWARDED

THE SECOND PRIZE OF THE BOYLSTON MEDICAL SOCIETY FOR 1867.

By F. R. STURGIS,

HOUSE SURGEON OF THE MASSACHUSETTS GENERAL HOSPITAL.

"SEGES VOTIS RESPONDET."

Printed for the Author.

CAMBRIDGE:
PRESS OF JOHN WILSON & SON.
1867.

PREFACE.

In presenting this Essay to the Society, I do not profess to bring forward any new ideas upon helminthic disease: it is nothing but a *resumé* of the latest ideas upon this subject, obtained from those works considered as authority upon such matters.

Among the treatises consulted are those of Küchenmeister, Von Siebold, Moquin-Tandon, Cobbold, Davaigne, &c., besides scattered communications from journals, which were deemed of interest.

Feeling the subject to be one of practical importance, as well as one of much interest, the writer feels, that, in presenting it to the notice of the Society, he has not worked for nothing.

The accompanying plates were copied from various works by a friend, Mr. H. P. Quincy, to whom I here desire to make my sincere acknowledgments.

January, 1867.

CONTENTS.

PART I.
DESCRIPTION, &c., OF TÆNIA SOLIUM 10

PART II.
DESCRIPTION, &c., OF BOTHRIOCEPHALUS LATUS. . . . 19

PART III.
DESCRIPTION, &c., OF CYSTICERCI, ECHINOCOCCI, AND ACEPHALOCYSTS 25

PART IV.
DIFFERENT SPECIES OF TÆNIÆ AND THEIR EMBRYOS . . 36

PART V.
DEVELOPMENT OF THE EMBRYO INTO THE TÆNIA . . . 60

PART VI.
PATHOLOGY AND TREATMENT 69

AN ESSAY

UPON

THE HUMAN CESTOIDS.

PART I.

BELONGING to the order of Sterelmintha, of the subclass Entozoa, these cestoids are capable of division into two genera; viz., Tænia and Bothriocephalus. Formerly a third was added, *i.e.*, Echinococcus; but, as this has lately been discovered to be a tænioid embryo, it is entitled to no generic distinction. The order of Sterelmintha is distinguished from the other order of this sub-class, the Cœlelmintha; inasmuch as, in its members, the intestine is in some cases wanting, in others vascular; while, in the second (the Cœlelmintha), the intestine is in a distinct cavity; the first being identical with the Vers Intestinaux Parenchymateux, the second with the Vers Intestinaux Cavitaires of Cuvier. It is my intention simply to review the parasites of the first order; to study the anatomy, embryology, &c., of the two genera; the pathological conditions they give rise to; and the best mode of treatment.

Tænia solium (Pl. I. fig. 1). *Description.* — These cestoids are flat and narrow, composed of articulations, slender, and but slightly flattened anteriorly; becoming broad and expanded as they recede from the head, and estimated of various lengths; 4 or 5 métres (12 to 15 feet), (Moquin-Tandon), never exceeding 24 feet (Weinland, from Diesing), from 9 to 35 feet (Aitken), and from 10 feet upwards (Owen); having, in this latter case, been found to extend from the pylorus to within 7 inches of the anus. The average length may be stated as from 10 to 12 feet. They are covered with a soft, nearly white integument, in which are scattered little granular masses, at one time supposed to be eggs, now proven to be carbonate of lime (Weinland *et alii*).

Head (Pl. I. fig. 2, B). — The head is of the size of a pin's head, and frequently colored with a black pigment; it is hemispherical or globate in form, and provided at the sides with four rounded nipples, diagonally placed; in the centre of each one is seen a circular disc which is imperforate (Pl. I. fig. 2, D; Küchenmeister). Anteriorly, and in the midst of these suckers, is noticed a convex protuberance, which is not perforated, and which is surrounded by a double row of hooks (Pl. I. fig. 2, E), from 22 to 30 in number (Moquin-Tandon). These little hooks are siliceous or cornuate in nature, and are composed of three portions, viz., the handle or root (Pl. I. fig. 2, C, a); the hook proper (Pl. I. fig. 2, C, γ), which is awl-shaped and pointed; and the guard (Pl. I. fig. 2, C, β), which is a protuberance situated at the middle third. These hooks are of various sizes, and are received into little sacs or follicles, which are disposed in a double row, as are the hooks themselves. These sacs have received the name of hook sacs (*Hakentasche;* Küchenmeister), and, when the Tæniæ become old, the hooks are lost; this loss is therefore considered as a sign of maturity (Von Siebold). As stated above, the cephalic protuberance is imperforate. At one time this was believed to be the mouth; this is now shown not to be the case: though whether these parasites are provided with an oval aperture or not, is yet an open question; some authors contending that they are supported by imbibition through the skin; others claiming the probable existence of a mouth, judging from the fact that it has been found in T. serrata and in T. osculata by Leuckart, and in Dibothrium claviceps by Wagener, in which species it was also supposed not to exist (Weinland). Passing downwards, we arrive at the neck (Pl. I. fig. 2, A), about 6''' in length (Küchenmeister), slender, slightly depressed, and transversely wrinkled, but not articulated. These wrinkles become more and more marked, and finally become transverse folds, dividing the worm into joints. These joints grow by a continuous pushing from the head and neck, whereby it follows that the joints nearest the head are the youngest (Weinland). They are not matured until they receive the sexual organs (Pl. I. fig. 3, A; and fig. 4), which first appear at the 280th segment, in the me dian line of the cestoid, as a simple brownish canal, with short lateral offshoots, towards which two transverse slightly colored lines (seminal cord and vagina) run from the sides (Küchenmeister). These are gradually developed in each succeeding joint, until, in the 650 to 700th segments, they are perfect, and filled with mature ova. The number of these joints have been estimated from 825 in a specimen 10 feet 8 inches in length (Küchenmeister), to 2,240 (Moquin-Tandon).

According to their position, these articulations have been found to differ in size, the first ones, those near the neck, being broader than they are long; afterwards, from being almost square, they become oblong; and finally their length is twice their breadth (Moquin-Tandon). These latter are narrow anteriorly, broad posteriorly, and overlap each succeeding one (Pl. I. figs. 4 and 1). The proportionate size of the different segments may be judged of from the following, copied from Küchenmeister: "First, a space of 4″ contained 50 transverse divisions, and afterwards this space showed 32, 27, 22, 14, 11, 10, 9, 8, 7, 6, 5, $2\frac{1}{3}$, 2, $1\frac{2}{3}$, 1, $\frac{4}{5}$, $\frac{5}{7}$, $\frac{2}{3}$, $\frac{1}{2}$ segments." As these joints ripen, they are cast off in the fæces, &c., containing within them the ripe fecundated ova. They are then known as proglottides (Vermes cucurbitini), and are capable, when cast off, of a considerable degree of locomotion, by alternate elongation and contraction of the muscular structure with which they are endowed; and this point has been adduced by Coulet ("Tractatus de Ascaridibus et Lumbrico lato"), in support of the idea that they are endowed with an independent existence. It has long been a subject of dispute, whether the proglottides should be considered as independent zoonites, derived from the Tænia, this latter therefore being regarded as a compound creation, formed of a colony of simple ones; or whether they should be considered as parts of a single individual. Weinland is of the opinion that the first is the true view; *i.e.*, that the whole tapeworm should be considered, not as one individual, but as a group of individuals, to be compared to the many individual Medusæ which are formed out of the Hydra fusca by transverse division. Von Siebold also entertains the same view of them, and considers their ability of locomotion, already spoken of, as a point in favor of it. These articulations are important parts of the parasite, containing, as they do, their sexual organization. Each one is androgynous; *i.e.*, containing in themselves the male and female genitalia united. The genital pore (Pl. I. fig. 4, A) is situated upon the margin of the joint, sometimes upon alternate sides, but often occurring for three or four times upon one side, and then upon the other (Moquin-Tandon). These first appear at the 317th joint, in the form of prominences; at the 350th, the pores themselves are distinct (Küchenmeister). This genital pore is very apparent, and opens at the summit of a mamillated projection (Moquin-Tandon). This pore, examined by the microscope, appears like a cup-shaped depression, with two small openings in it, through the anterior one of which the lemniscus or rudimental penis projects (Pl. I. fig. 3, *a*). This organ, which is about 0·276 mil. in length,

0·071 mil. in breadth behind, and 0·395 mil. in front, is yellowish, sickle-shaped, muscular, and covered with bristles (Küchenmeister, Weinland). This is received into a preputial covering, which is 0·175 mil. in length (Küchenmeister); from the posterior portion of which runs the vas deferens (Pl. I. fig. 3, A, β), which is twisted, of an opaque white, and leads to the testicle (Pl. I. fig. 3, A, γ) situated in the middle of the articulation, and at right angles to the uterus. Küchenmeister says he has never been able to trace a testis; but in this connection he speaks of the cord " which takes a tortuous course, running at a right angle towards the uterus, and running nearly to this, in which it forms convolutions of 0·031 mil." Now, as I understand it, this convolution is the testis, which is thus formed in nearly all of the animal kingdom. It surely cannot be the vas deferens?

The female genitalia (Pl. I. fig. 3) are more difficult and complicated in their structure. Posteriorly to the male genital opening is another one (Pl. I. fig. 3, A, δ), with a funnel-shaped aperture of 0·063 mil. in breadth (Küchenmeister). This is the vulva, and communicates with a canal, which first runs parallel with the vas deferens; but, quitting this in the middle of its course, winds tortuously into a straight, granular canal (Pl. I. fig. 3, A, ϵ), the uterus, into which open dendriform, multilobular branches, from 9 to 15 in number (Küchenmeister), which alternate irregularly, and resemble somewhat a bunch of grapes (ovaries; Pl. I. fig. 3, A, η).

I would suggest, that this median stem, instead of

being called the uterus, be considered as a part of the ovaries. It is apparently of the same texture as the ovaries; its functions appear the same. Why then call it a uterus, when it is not destined to contain the embryo at all, but is continually filled with eggs, which are generally more developed here than in its branches? I would therefore consider it as a portion of the ovary, not as a separate organ, and look upon the Tænia as destitute of any true uterus.

No visual or respiratory organs have been discovered in this parasite; and it seems, *a priori*, that there should be none, as it lives in closed cavities or dark recesses, where there would be no occasion for them.

Starting from the sucking discs are seen four little filiform canals. These unite and form two, one on each side, which run down along the margins of the entozoon; uniting at the posterior border of each joint by transverse canals. These lateral canals possess valves, which prevent their fluid contents from regurgitating (Moquin-Tandon). Weinland does not consider that they unite, but remain separate; all four finally meeting in the head of the worm, in a vascular ring, round the proboscis.

This has been considered as the nutritive system of the Tænia; its contents consist of a watery fluid (Weinland).

No anal orifice has been observed.

With regard to the circulatory apparatus, M. Blanchard described four thin longitudinal canals, two on each aspect, communicating by means of four very fine vessels. M. Van Beneden, of Belgium, considers them as urinary vessels. Is it not possible that they may have been confounded with the nutritive system? The description answers well.

The nervous system of these parasites is a subject of interest and much doubt, being described as consisting of two cerebral ganglia united by a slender commissure, from which long cords go off upon each side of the articulations (nutritive system?). It has also been stated, that at the base of each disc is a little ganglion, which communicates with the cerebral ones. M. Dujardin does not view these lateral cords as nerves, but rather as ligaments (Moquin-Tandon).

The skin is soft, nearly white, rather tough, and consists of a corium, thick and fibrous, and of an epidermis. Beneath these there are muscular fibres, both longitudinal and transverse (Küchenmeister), which are not interrupted at the borders of the articulations.

We have already seen that the ova are contained in the older joints only (about the 700th), and that those near the posterior end of the entozoon are the ones which are fecundated. This is performed by the intromission of the male organ into the female opening.

The eggs (Pl. I. fig. 3, B) which are thus fecundated are 0·036 mil. in diameter, round, and rather whitish, and consist of two or more shells, the outermost of which is chitinous, either thick and composed of many small granules, or thin and membranaceous. Only those which are matured possess the two shells;

the immature ones (those contained in about the 600th articulation) having but one (Weinland). The outer covering is very tough, and resists the action of dilute acids, alkalies, and exposure to heat, cold, dampness, or dryness (Aitken). The reason of this is evident at a glance. As the embryo is hatched outside of the body of the parent entozoon, it would be much exposed were it not for the resistance offered by its habitation to external influences.

The number of these eggs in a single proglottis is almost incredible, being counted by thousands.

Within the covering may be seen the embryo (Pl. I. fig. 1), $\frac{1}{1250}$ of an inch in breadth (Cobbold), of globular form, and provided with six spines arranged in pairs, one on the top and one on each side, a little anterior to the middle.

The development of the embryo in the egg appears to take place by segmentation, as may be seen in Pl. II. fig. 2, after Leuckart (copied from Aitken's "Science and Practice of Medicine," vol. i.).

Habitat. — These entozoa are found in the small intestine; but they may grow to such a size as to extend into the large, as in the case quoted from Owen, in the beginning of this chapter. In rare cases, they mount into the stomach. Aubert, of Geneva, has described a tumor of the testicle, caused by a Tænia (Moquin-Tandon). These are very rare, even if not very doubtful, cases, and do not prevent their being considered as inhabitants of the intestine. They are found attached to its mucous membrane by their hooks, and are stated to be capable of undulatory movements (Gomés, Deslandes).

Geography.—This parasite is found more frequently in France, Italy, Holland, Germany, Great Britain, Egypt, and Abyssinia, than in other countries; and, although there seems no reason to suppose that the conditions for their growth may not exist in other places, the cause may probably be found in the difference of food. The Bothriocephalus latus (next to be described), the embryo of which is supposed to live in fishes, occurs in Russia, Sweden, Norway, Lapland, Finland, Poland, and Switzerland, most of them countries bordering on the sea-coast, and using fish and vegetables in larger quantities than those others above mentioned.

In former times it was supposed that these parasites existed singly in the intestine, hence the name. This is a fallacy; and it is now known that several may reside together in the same intestine.

PART II.

Bothriocephalus latus.

BELONGING to the same order as the previous entozoon, the genus Bothriocephalus differs in so many respects from it, as to be worthy of a separate description. I have therefore introduced it here, for convenience sake, before considering the different species of Tæniæ.

Description. — This parasite (Pl. I. fig. 5), first described by Bremser, is also articulated, of a bluish white or gray color, of a length varying from 2 to 7 mètres (6 to 22 feet), and a breadth of from 10 to 13 mil. (Moquin-Tandon).

Its head (Pl. I. fig. 6) differs largely from that of a Tænia, being small (about 2 to $2\frac{1}{2}$ mil. long), oblong or ellipsoid in shape, and retracted in front (Moquin-Tandon). In the perfect state, it is devoid of hooks, such as are found in the Tæniæ; and its suckers are of a different shape, being lateral pits or depressions upon opposite sides of the head, instead of circular ones (Pl. I. fig. 6, A). These depressions are not bounded by any borders.

As in the Tænia, no mouth has been discovered, although Bremser is of the opinion that it is situated between the two cephalic depressions. Later investigations have not supported this opinion; and the belief now is, that they are probably nourished by imbibition through the skin.

Beneath the head is the neck (Pl. I. fig. 6, B), varying in length (being sometimes developed, and sometimes not), and does not possess articulations, although by the microscope quite marked transverse striations are visible. The articulations (Pl. II. fig. 3), which, according to Aitken, commence about three inches from the head, likewise differ from those of the Tæniæ. At first being square, they become broader than they are long, generally in the proportion of 3 : 1 (Küchenmeister); although, even here, they become capable of variation. Each articulation has two surfaces, — a ventral (upon which are situated the genitals) and a dorsal one; and four margins, two lateral, which are undulatory; and one anterior and one posterior, by which it is joined to the segment preceding and succeeding. As in the Tæniæ, the anterior articulations overlap the posterior ones, but not in so marked a degree.

The number of articulations which an adult Bothriocephalus bears, has been reckoned by M. Eschricht at 10,000 (Moquin-Tandon). In the centre they are thicker than at the sides, up to 1″ (Küchenmeister), and possess a dark-brown color, according as they contain matured ova, becoming at the margins flatter and whiter.

The genital organs in this genus are situated, not, as in the Tæniæ, on the margins, but in the centre of the articulations, and are recognizable as slight elevations (Pl. II. fig. 3, A). In some articulations, both sexes exist; in others, sometimes the male, and sometimes the female, separately.

The male opening (Pl. II. fig. 3, B) exists about the middle of the articulation; the female (if both sexes be present together) below it, and near the posterior margin (Pl. II. fig. 3, C). The penis (Pl. II. fig. 4, a), of equal thickness throughout, $\frac{1}{8}'''$ (Küchenmeister), is easily protruded about $\frac{1}{4}'''$ beyond the margin of the opening. The prepuce is formed by this margin of skin, which is undulatory in character, and in which are scattered calcareous corpuscles, — the "glandulæ preputii" of Eschricht. The penis communicates with a pretty long deferent canal, several times wound upon itself, which gradually increases in thickness, and terminates in a vesicula seminalis in the form of an obovate pouch (Moquin-Tandon). Dr. Küchenmeister, in his work, " Die in, und an, dem Körper des lebenden Menschen vorkommenden Parasiten," Leipzig, 1855, does not speak of a vesicula seminalis; but, under the name of sac of the penis, describes " a vesicle situated in a proper capsule, $\frac{1}{4}'''$ in length, $\frac{1}{8}'''$ in breadth, between the uppermost thick horns of the uterus, and in which the penis is suspended to a strong twisted stem, $\frac{1}{4}'''$ in length," which I presume to be the analogue of the vesicula seminalis described by Moquin-Tandon. Leading from this, and placed between the lateral uterine horns upon its dorsal aspect, are several fine delicate canals, from $\frac{1}{6}$ to $\frac{1}{25}'''$ in thickness (Küchenmeister), which, according to Moquin-Tandon, go to the testis.

Küchenmeister, however, denies the communication, and states that they have "no perceptible connection with the vesicle of the penis, or with the testis."

These latter (Pl. II. fig. 4, β) are composed of whitish granulations, 0·030 to 0·080" in size (Küchenmeister), and are situated at the uppermost portion of the genital apparatus. They contain, according to Eschricht, fine, spiral filaments (spermatozöoid filaments; Küchenmeister).

Below the male orifice, and at the lower part of the articulations, is the female opening, consisting of the vulva (Pl. II. fig. 4, γ), which leads into a tortuous canal, the oviduct (Pl. II, fig. 4, δ); while on either side are several horns or pockets, which Küchenmeister designates as the uterus, but which appear to me more likely to be the ovaries (Pl. II. fig. 4, $\varepsilon, \varepsilon, \varepsilon, \varepsilon$).

A little below the female opening, and on either side of it, are two other horns, united together by a canal; this M. Moquin-Tandon considers as the uterus (Pl. II. fig. 4, η, η). I am inclined also to regard these as ovaries, and to deny the Bothriocephalus a uterus, for the same reason that I did the Tænia. Küchenmeister, besides, describes a pigment capsule (Eschricht's coil), which secretes the brown coloring matter, and perhaps the shells of the ova; it is 5 or 6" in length, enlarged like a sac in the middle, and opening into the uterus directly at its commencement.

As might be expected, these entozoa possess no visual or respiratory organs; at least, none have ever been observed.

As in the Tæniæ, there are seen, in the anterior portion of the body, two filiform longitudinal canals, filled with a limpid fluid. These Moquin-Tandon considers to be the alimentary canal. Küchenmeister, who observed the same appearances, does not give any opinion as to what they may be.

"According to M. Blanchard, this parasite possesses a nervous system, analogous to the Tæniæ, but much less distinct" (Moquin-Tandon). Küchenmeister, however, states in his work (*op. cit.*), that no nervous system has been discovered.

The skin presents nothing very worthy of remark. It is divided, as in the Tæniæ, into a corium and epidermis, which is inverted at the generative openings, and which contain calcareous corpuscles; more abundant about the genital openings, and the hinder margins of the segments. The parenchymatous portion shows contractile cellular tissue, which consists of threads, running at right angles, forming a wide network, containing within them calcareous bodies; and also very weak transverse and stronger longitudinal muscular striæ, which, however, are less dense than in the Tæniæ (Küchenmeister).

The eggs of this species (Pl. II. figs. 4, B, 5) are quite large, of an elliptical or ovoid shape, of a brown color, and consist of but one covering (Weinland). which is hard, brittle, and translucent, and through which the segmented yolk is visible. They are furnished with an operculum (Pl. II. fig. 5, *a*), and are very numerous; each individual Bothriocephalus, according to M. Eschricht, containing ten millions of eggs (Moquin-Tandon).

These ova contain the embryo, which at first is oval, afterwards becoming globular, is enclosed in a ciliated covering, and furnished with six hooks (Pl. II. fig. 5, A).

Habitat. — The location of these parasites, like the Tæniæ, is in the small intestines.

Geography. — They are more frequently found in the north of Europe, viz., Russia, Sweden, Finland, &c., than in other parts.

PART III.

Cysticerci, &c.

HAVING discussed the two preceding entozoa, there yet remains another group to mention, viz., the Cysticerci. The true nature of this group, and the relation they bore to the Tæniæ, were for a long time unknown; and, although the resemblance between the heads of some of the Cysticerci and of the Tæniæ had been noticed, no practical deductions were made therefrom for some time.

Weber, in 1688, was the first who discovered the similarity existing between the C. fasciolaris and the tapeworms.

In 1760, Pallas described these Cysticerci, as forms of tapeworms, under the name of Tænia hydatigena.

In 1782, Ephraim Goeze, a German clergyman, in his work, "Versuch einer Naturgeschichte der Eingeweiderwürmer," noticed the similarity between the heads of the C. fasciolaris, found in the liver of the mouse, and the T. crassicollis (his T. serrata) of the cat, as is shown in these words (*op. cit.*, p. 340), while speaking of the T. crassicollis:—

"The size, form, and structure of its head are perfectly identical with those of the head of the articulated cystic tapeworms in the liver of the mouse; for this also has no neck, but its head sits immediately upon the first segment." And, while speaking of the C. fasciolaris of the mouse, he says,—

" The size of its head agrees perfectly with that of the tapeworm with notched segments, (T. serrata)" (Küchenmeister).

That Goeze, moreover, had some idea, that from the C. fasciolaris was produced the T. crassicollis, may be gathered from the following (*op. cit.*, p. 245): " On the 13th March, 1780, I found, in the liver of the mouse, two clear crystal vesicles, in each of which there was a pisiform vesicle, but on this as yet no body. I believe, that, as regards the production and development of this kind of worm, I have surprised Nature in the act. In the interior of the inner vesicle, there was a small white process or body of about 1″ in length. This was firmly attached by its base to the interior of the vesicle, and the white point at which it was affixed could be seen from the outside. When the vesicle was placed in such a position that the white point to which it was attached was at the bottom, it stood upright, in the interior of the vesicle, like the light in a lantern, reaching about to the middle of the vesicle, so that it was completely surrounded by the vesicle." From this it follows:—

" 1°. That this was the first stage of growth of the cystic tapeworm in its vesicle.

" 2°. The first thing that comes out of the egg must, therefore, be the caudal vesicle; and this because the worm must first care for its habitation, and prepare this in proportion to the growth of its caudal vesicle.

" 3°. In the vesicle, the body sits, but internally, and, as it were, turned inside out. It must therefore live upon its own juices in the vesicle, until it is time to reverse itself; because it already has the four suckers and the circlet of hooks upon its head. The caudal vesicle thus seems as a reservoir of nourishment.

" 4°. When its body has attained the necessary degree of development, and the vesicle over it is no longer large enough for its habitation, the body reverses itself, by the agency of its folds and segments, from within outwards, and then constantly grows until it reaches its perfect form and size, such as we procure it from the cysts of the liver.

" 5°. The body here sits still in the vesicle, exactly in the same way that the numerous bodies of the cystic tapeworm of the sheep sit in the interior of their common vesicle, in the manner of a colony " (Küchenmeister).

Still the development was not followed out until Steenstrup, of Denmark, by his " Theory of the Alternation of Generation," in 1842, supposed that these cystic worms were early steps in the development of the generation of Helminthes. Dujardin, of France, in 1845, in his " Histoire Naturelle des Helminthes," pp. 554, 633, and Von Siebold, of Germany, in the second volume of Rud. Wagner's " Handbuch der Physiologie," art. *Parasiten*, were the first who stated that these cystic worms were undeveloped animal forms, and young states of tapeworms ; and, indeed, that they were produced from those germs of tapeworms, which, instead of the intestine, had got into the parenchyma of their host, and, under the influence of this unusual dwelling-place, had advanced to the abnormal state of development which we call a cystic worm.

In 1850, Van Beneden, of Belgium, in his work, "Les Vers Cestöids où Acotyles," considered these cystic worms to be the larval condition of the Tæniæ; but, in his further conclusions, he made the mistakes, 1st, of considering that a tapeworm can pass through all the separate phases of its development in the intestine of its host; and, 2d, in considering that the caudal vesicle is formed by dropsical degeneration, and that the head there sinks into it in order to become a Cysticercus (Küchenmeister).

In 1851, Küchenmeister, from some experiments with the C. pisiformis of the rabbit, and the C. fasciolaris of the rat, arrived, among other conclusions, to these two, that, —

1°. "The cystic worms are not strayed dropsical tapeworm nurses, but tapeworm larvæ furnished with a provisional organ (caudal vesicle), probably acting as a reservoir of nourishment, and incapable of sexual multiplication, for which there is neither room nor sufficient nourishment;" and that,

2°. "The cystic worms constitute a necessary step in the development of the Tæniæ."

Furthermore he says, —

"At the same time I proved that the Cysticerci administered, when transferred to the intestine of other animals (as for instance the C. pisiformis in the intestine of the cat, &c.), did not become developed into jointed tapeworms, but bore behind the head a long, inarticulated appendage, and died in a short time without being further developed; and that every species only thrives in a particular species of animal."

In the following year, Von Siebold repeated these experiments, and in the main confirmed those of Küchenmeister; differing from him, however, in his last conclusion.

I shall consider these parasites, therefore, not as a separate group, but, as they are now understood to be, the undeveloped or embryonic conditions of the Tæniæ; and, as an example of these embryos, I shall select the Cysticercus cellulosæ.

These Cysticerci (Pl. II. fig. 6) appear as small, elliptical, or rounded bladders, fibrous in structure, and are contained in the tissue of the organ where it has taken up its abode. This bladder or cyst (Pl. II. fig. 6, A) is resistant to the knife; and when opened it is found, that, by pressure, another vesicle can be squeezed out. This vesicle contains an opening, around which is adherent a third sac, which is of globular, pyriform, or oval shape, semi-transparent, of whitish color, with quite thin, fine, granular walls, and but little resistant. These cysts in the C. cellulosæ present a long diameter of 15 to 20 mil. and a short one of from 5 to 6 mil. Within the cyst is contained the scolex, as it were, doubled upon itself; *i.e.*, the head and neck retracted upon itself. When in this position, its site upon the outside sac is marked by a little umbilicated depression surrounded by a more or less white sphincter (Moquin-Tandon; Pl. II. fig. 6, A, a).

When unfolded (Pl. II. fig. 6, B), this Cysticercus is seen to possess a head (Pl. II. fig. 6, B, a), a neck (Pl. II. fig. 6, B, b), and then to terminate in a sac filled with a fluid containing fat, albumen, and calcareous matter (Küchenmeister). This sac is the caudal vesicle (Pl. II. fig. 6, B, c). In this caudal vesicle, which is identical with the third sac spoken of above, the parasite is contained when in a folded condition.

The head and neck are very similar to those of the Tæniæ; and this resemblance had been noticed long before the connection between the two had been established. The head is furnished with sucking discs (Pl. II. fig. 6, C, a), from four to six in number, round which the vascular system runs, and afterwards collects into two longitudinal canals on each side (Küchenmeister); and is surmounted by a rostellum or proboscis, surrounded by a circlet of hooks varying in number from 22 to 32 (Pl. II. fig. 6, C, b). This head is colored with a yellowish or blackish brown pigment; and there are found, round the stem of the hooks, five sacs (Küchenmeister).

The neck (Pl. II. fig. 6, C, c), of varying length, is rugous, as it is in the Tæniæ, like them not articulated, and contains a few calcareous corpuscles.

The caudal vesicle (Pl. II. fig. 6, B, c) is formed of contractile tissue, and contains circular parallel rings. It is homogeneous, destitute of vessels and calcareous corpuscles, and consists of an organic substance belonging to the class of mixed protein substances, and is nearly allied to chitine, but dissolves, when boiled in caustic alkalies, rather more easily than the latter. No re-action of cellulose follows upon the addition of iodine (Küchenmeister).

This form corresponds to the second stage of the developmental process of the parasite, the scolex.

Habitat.— Its habitation is various, being found in the muscles (Werner *et alii*), in the liver (Dupuytren, Leuckart, Laennec), in the heart (Morgagni, Rudolphi, Bouillaud, Andral, *et alii*), in the choroid plexus (Treutler, Brera, Fischer), in the brain (Ruysch, Chomel, *et alii*), in the anterior and posterior chambers of the eye (Sœmmering, Von Græfe), in the vitreous humor; in and under the retina (Von Græfe), between the sclerotic and conjunctiva (Von Siebold *et alii*), and in one instance in bone, the first phalanx of index finger (Stanley). When in the closed cavities of the body, *e.g.*, in the ventricles of the brain and in the eye, the Cysticercus lives free; in other parts of the body, it is surrounded by a cyst, formed from the tissues of the body in which the parasite takes up its abode.

Echinococci.

We now pass to the consideration of the second class, viz., the Echinococci (Pl. III. figs. 1 and 2). These parasites are contained in a capsule (the hydatid cyst of authors), varying in size from a grain of mustard to that of a hen's egg or more, and of a globular, ovoid, or pyriform shape. Within this external capsule is another one closely united to it, and containing the scolices; both of them again being contained in a tough, fibrous covering, variously colored, according to the organ in which it is placed, closely adherent to the surrounding tissue, abundantly supplied with blood-vessels, and sending out into the organ where it is found prolongations of connective tissue.

The external covering is of an albuminoid consistence, laminated, containing no fibres, fibrils, or cellules (Davaigne). The internal vesicle (the germinal membrane of Goodsir) is a fibrillated, granular structure, containing a vascular system and calcareous corpuscles (Küchenmeister). In its interior is found an almost colorless, transparent, limpid fluid (Moquin-Tandon). It is upon the interior of this membrane that the Echinococci are formed.

This takes place in the following manner:—

From the internal vascular layer of the Echinococcus vesicle spring small buds, which form conical or villous-like elevations, sometimes measuring 0·4 mil. These become transformed into small brood capsules 0·07 to 2 mil. in diameter, and in these the scolices or buds are formed. This process, which at first adheres by a tolerably broad base, becomes in this way cleared, its contents more fluid; and a small globular vesicle, adhering to the inner wall of the Echinococcus, is found, presenting an external structureless epidermis and an inner granular layer. In the granular layer of these processes, which have been dilated into vesicles, we observe vessels, which are connected with those of the Echinococcus vesicle, and subsequently, when these small vesicles have attained a certain size, again lead to the formation of processes in themselves. Thus, from these new (4 to 10) processes, the scolices of the Tæniæ are produced, with a general increase in the size of the capsule. The wall of the vesicle, in which the head is inverted, afterwards forms the hinder part of the young Tæniæ. When the Echinococcus heads are developed in the interior of the process which has become converted into a vesicle, which always takes place uniformly, the brood capsule bursts, and its walls, and with them the inner layer, with the individual heads, turn inside out (Küchenmeister).

The anterior portion or head (Pl. III. fig. 1, A, a) of these embryo parasites are formed as are those of the Tænia or Cysticercus. They possess the sucking discs (Pl. III. fig. 1, B, a) and a double circlet of hooks, varying from 46 to 54 in number (Pl. III. fig. 1, B, b; and fig. 3).

This is followed by the neck, which is short, and contains calcareous corpuscles (Pl. III. fig. 1, B, c).

The body is discoid rather than globular, also filled with granular corpuscles, sometimes attached to the membrane by a pedicle (Pl. III. fig. 1, A, b), and sometimes not, and is $\frac{1}{20}$ to $\frac{1}{10}$ of a line in length.

The hooks (Pl. III. fig. 1, C) of this parasite are, according to Eschricht, 0·02 to 0·022 mil. in length.

In neither the Echinococcus nor the Cysticercus have any sexual organs been found.

Habitat. — These parasites are found pretty generally distributed over the entire body; viz., in the liver (Rudolphi, Eschricht, Lebert), the lung, the kidney, the brain (Zeder), in the lachrymal gland (Smidt), and in the heart (Morgagni). They have also been found in the spleen (Collet), in a goitre (Albers and Boeck, by Förster), in the choroid, and in the crystalline lens (Geschiedt).

Acephalocysts.

The third group, those of Acephalocysts (Pl. III. fig. 4), are now considered to be a barren Cysticercus or Echinococcus vesicle, in opposition to the former belief, that they were a separate species of Helminth, or that they were not of an independent animal organization.

They are vascular, of a spherical shape, of simple structure, non-adherent to the tissues, and containing within them other vesicles, attached to the inner wall of the external vesicle or to each other. The walls are gelatiniform, elastic, and tremble, when touched, like a jelly. This vesicle contains a watery fluid, or a substance which has a purulent consistence, and contains the microscopic elements of calcifying embryos, or of encysted proteine masses in the act of resorption (Küchenmeister). Within the secondary cysts are, according to Küchenmeister, no solices of cestoidea or their remains to be found: but Van Beneden, of Belgium, does not agree to this; for he states that it is not rare to find sacs containing at the same time inert vesicles, and embryos provided with their hooks, which consequently pertain to the nature of Acephalocysts or Echinococci. They are now, therefore, considered as six-hooked embryos (cestoid), the growth of which has proceeded without hinderance, but which, nevertheless, have remained barren, or, more correctly, which have never attained to proliferation and the production of scolices (Küchenmeister).

PART IV.

Different Species of Tæniæ, &c.

In the preceding parts, I have hurriedly described the different genera of the Cestoids of man and their embryos, selecting as the types the most common ones. It now remains for me to describe the different species belonging to each. They may be arranged thus:—

Of the Tæniæ.

1. Tænia solium (Linnæus).
2. Tænia solium. Varietas abietina (Weinland).
3. Tænia mediocanellata (Küchenmeister), or Tænia inermis (Moquin-Tandon).
4. Tænia nana (Von Siebold). Tænia ægyptica (Bilharz). Scolex unknown.
5. Tænia (Hymenolepis) flavopunctata (Weinland). Scolex unknown.
6. Tænia capiensis (Küchenmeister).
7. Tænia echinococcus (Von Siebold, Küchenmeister).
8. Tænia cucumerina (Bloch). T. elliptica (Batch), or T. canina (Pallas). Scolex unknown.
9. Tænia lophosoma (Cobbold). Scolex unknown.

Of the Bothriocephali.

1. Bothriocephalus latus (Bremser).
2. Bothriocephalus cordatus (Leuckart). Scolex unknown.

Of the Cysticerci.

1. Cysticercus tæniæ cellulosæ (Rudolphi), or C. cellulosæ. Scolex of Tænia solium.
2. Cysticercus tæniæ mediocanellatæ (Leuckart). Scolex of Tænia mediocanellata.
3. Cysticercus tenuicollis (Rudolphi). Scolex of Tænia marginata (the mature form is common in the dog); not found in man.
4. Cysticercus acanthotrias (Weinland). Tænia unknown.

Of the Echinococci.

1. Echinococcus hominis (Rudolphi). Echinococcus altricipariens of Küchenmeister. Tænia unknown.
2. Echinococcus veterinorum (Von Siebold). Echinococcus scolicipariens (Küchenmeister). Scolex of Tænia echinococcus.

And of the Acephalocysts.

1. Those derived from the Tænia echinococcus altricipariens of Küchenmeister.
2. Those derived from the Tænia echinococcus scolicipariens of Küchenmeister.
3. And those derived from the Tænia ex Cysticerco tenuicolli (Küchenmeister).

The first of this group, T. solium, has already been described: it will therefore not be necessary for me again to speak of it.

The Varietas abietina of the Tænia solium, so called from the resemblance which the uterus bears to a pine-tree (abies) is described by Weinland, in his essay on the "Tapeworms of Man." It was obtained from a Chippewa Indian, at the Saut Ste. Marie at Lake Superior, by Professor Louis Agassiz:—

"The specimen consists of a chain of several feet in length, from the mature part of the worm. The head, neck, and whole anterior half, are wanting. The most striking thing in this worm is its extreme narrowness and meagreness, while Tænia mediocancellata, which it resembles in the configuration of its uterus, is very broad and thick, according to Küchenmeister. All the joints which are preserved are very thin, nearly transparent, and equally narrow; their transverse diameter being about 4 mil., and the longitudinal about 12 mil. The genital openings are very small, and without external lips: this may be owing to the very mature age of the joints in question. There is no pigment in either vagina or spermatic duct. The uterus is more regular than either in T. solium or T. mediocanellata, yet it more resembles the latter. The middle trunk of the uterus is quite straight; the branches, about thirty in number, start from the main trunk, either at a right angle, or at an angle of about 45°. These branches are always quite parallel, and are generally straight; but, whenever they are bent, they all make the same angle. They are never arborescently divided, nor furcated at the ends, with the exception of the foremost and the hindmost in each joint, which run, the former forwards, the latter backwards; both being forked and crooked. The eggs, which are extremely plenty in these joints, and which show the configuration of the uterus in a yellowish tint to the naked eye, are 0·033 mil. long and 0·030 mil. broad. They are protected, first, by an outside shell (Chorion), which is 0·003 mil. thick, dark in its outside layers, transparent, yellowish inwards; then follows a second shell (yolk membrane), 0·0006 mil. thick, entirely transparent. In the cavity of the egg lies the embryo, occupying about two-thirds of it, and measuring only 0·016 mil. We saw other eggs unripe, and with one egg-shell only, but very rarely."

The second on the list, the Tænia mediocanellata (Pl. I. fig. 7), was first described, as separate from Tænia solium, by Küchenmeister, who in all saw seven heads. The body of this species is broader than that of the Tænia solium: the epidermis is thick, very distinct, soft, and without calcareous corpuscles. Beneath this is a layer of longitudinal muscles, running the whole length of the body, which form bundles of 0·545 mil. Then, below, follows a layer of transverse muscles, in which, as well as in the layer before it, are found calcareous corpuscles. The head is larger than in the Tænia solium, as are also the sucking discs, which are four in number, and of a black color (Pl. I. fig. 7, a). In this species, the rostellum and circlet of hooks are entirely wanting, in consequence of which Dr. Weinland has suggested the name of Tæniarhyncus (ταινία, a privitive, and ῥύγχος, a muzzle). At first this deficiency of the hooks gave rise to the question, whether these might not be T. solium, which had lost their circlet of hooks by old age, &c., instead of a new genus of Helminths. This question has, I think, been sufficiently settled by the experiments of Leuckart and Mosler in Germany, and Simonds and Cobbold in England, with the scolices of this Tænia, of which I will speak when I consider the scolex of this entozoon.

The vascular network consists of a transverse branch, running through the free space between the four sucking discs (Pl. I. fig. 7, b); from this, a branch runs to and around the sucking discs, until the four well-developed longitudinal vessels are seen running from them in the neck (Pl. I. fig. 7, c, c). No anastomoses are discovered between the individual branches of the sucking discs; neither is there any anastomosing transverse branch on the posterior margin of each segment, as are found in the Tæniæ. At these points, however, are found small enlargements of the vessels, distinguished by a kind of valvular apparatus, which appears to open before the fluid streaming from the head, but closes itself against that running towards the head.

In the shape of the segments also, there is a difference between this species and the Tænia solium. At first they have a tendency to increase in breadth, being about 6 mil. broad by 1 mil. long. This continues for some time, the articulation becoming 10, 14, 15, 17 mil. broad, by from 9 to 14 mil. long. This proportion does not remain constant, and then segments are seen from 3 to 4 lines broad by 1 to 1½ inches long. This change of form Küchenmeister explains by the fact, that only the longitudinal muscles reach from the upper to the lower margin of the segments, while the transverse muscles cease at a greater or less distance from the margins.

The genital openings, unilaterally placed, are large and swollen. The penis is thicker and shorter than is that of the Tænia solium; is 0·316 mil. long, 0·031 mil. broad at its apex, and 0·063 mil. at its base, and leads posteriorly into a convoluted seminal cord, from 0·023 to 0·039 mil thick.

Beneath this is the female genital orifice, strongly marked with pigment, 0·071 mil. at its external orifice, then diminishing to 0·039 mil. in its course; and opens, in the lower third of the segment, into the uterus with a dilatation of 0.079 mil., running first along the lower side and parallel to the seminal cord, until it suddenly bends downwards.

The uterus is a thick-walled, straight, median canal, with numerous lateral branches, which spring opposite to one another, and run parallel, and perfectly undivided, into the margins of the segments, when they either terminate in a cæcum, or, at the utmost, become furcated; but do not, as in the Tæniæ, divide dendritically.

The ova, 0·036 mil. long, and from 0·014 to 0·033

mil. broad, are usually rather smaller, lighter brown in color, smoother, more oval, and less globular than are those of the Tænia solium. The egg-capsules show only two consecutive layers, and are more easily broken than are those of T. solium (Küchenmeister).

4. *Tænia nana.* — This entozoon was discovered (and but once seen) in 1851, in Cairo, Egypt, by Dr. Bilharz, in large quantity, in the small intestine of a young man, who died of meningitis. It is of small size and slender, having a length of only 13 mil. (Moquin-Tandon), and is described by Küchenmeister as having broad and well-developed segments, and a large, quadrangular head (flat in front, and gradually diminishing in breadth), at the angles of which round discs are placed upon globular elevations. The rostellum is pyriform, and surrounded by a double row of small hooks. No measure of these latter have been given. The neck is slender, and is followed by segments, which gradually become broader, until, at the hinder end of the body, they acquire three or four times the width of the head.

I can find no description of its sexual organs; its eggs are globular, with a thick yellowish capsule, and beneath that a thin, vitelline membrane (Bilharz), within which, in fresh specimens, the six hooklets of the Tænioid embryos may be seen.

5. (*Tænia*) *Hymenolepis flavopunctata* was first described by Weinland, of Frankfort, while in America, in 1858. It was presented to the Society for Medical Improvement, of Boston, by Dr. Ezra Palmer of that city, in 1842. There were in the phial six different

specimens of this entozoon; none of them, however, perfect. The parasite was obtained from a "healthy infant, nineteen months old, which had been weaned about six months, and had had the usual diet from that time. The worm was discharged without medicine, its presence having never been suspected."

Owing probably to the regularity and shortness of the joints, and the presence of a yellowish spot situated about the middle of each joint, it was catalogued as a Bothriocephalus until Dr. Weinland described it as a new species of Tænia.

I quote his description of it: "The length of the whole worm is between 200 and 300 mil.; that is, from 8 to 12 inches. There were pieces of 50 mil. in length, consisting of very young joints, only $\frac{1}{5}$ mil. long, and 1 to $1\frac{1}{2}$ mil. broad; again, other pieces, about 100 mil. long, consisting, in their anterior half, of white, immature joints, $\frac{1}{3}$ to $\frac{1}{2}$ mil. long, and $1\frac{1}{2}$ to 2 mil. broad, while the mature joints of the posterior half, which are of a greyish tint (produced by the eggs they contain), average 1 mil. in length, and $1\frac{1}{2}$ to 2 mil. in breadth. In the young joints, the sides form straight lines, the transverse diameter being equal throughout the joint; in the riper ones, they are round and bulged, and the transverse diameter is the greatest in the midst of each joint. One of the pieces which is especially mentioned in the Catalogue of Dr. J. B. S. Jackson, shows the form of the joints when fully matured, and soon to be freed as proglottides. They are in this specimen triangular in shape, being narrow in front and suddenly broadening behind, evidently having already discharged the eggs from the anterior part of the joint; while, generally, proglottides deposit their eggs only after they are free. In other specimens, these last joints, being yet quite full of eggs, are more oblong, even with the transverse diameter longer than the longitudinal. In either case, the proglottides are very loosely connected with each other.

"In relation to the genital organs, we have mentioned above the yellowish spot lying near the middle line in the anterior part of each joint, and it is for this that we have called the species 'flavopunctata.' These spots are the testicles, appearing under the microscope as a globular gland, with another smaller one attached to it; this latter one runs out towards the side of the joint, into a longitudinal canal, in which lies the penis.

"The genital openings are situated all on one and the same side of the worm, while, in all true Tæniæ known thus far, they are found irregularly, now on one, now on the other side.

"The configuration of the uterus, also, differs greatly from that in the genuine Tæniæ. There is no main stem in the midst, with lateral branches, as in the latter; but, on the contrary, the eggs are crowded over the whole joint. It sometimes appears as if they were arranged in straight lines along the joint; but this is certainly owing only to regular lines of muscular contractions. Only fresh specimens can decide ultimately the structures of the uterus. From a careful dissection of the younger joints, we should judge that it consists of globular blind sacs, located here and there in the joint, and connected by fine tubes, terminating finally in the vagina.

"The most characteristic feature in this worm is its eggs, the number of which may be counted by thousands in each ripe joint. They are very large, measuring 0·054 mil. in diameter, and, under a low power of the microscope, appear as transparent balls, with a yellow dot in them. With a higher power, we easily distinguish three distinct egg-shells. The outside shell is translucent, elastic, cracking in sharp angles under pressure, and only 0·0007 mil. thick; this shell is folded under application of glycerine. The second shell is membranaceous and irregularly wrinkled, thinner than the first, and immediately attached to it. This second shell showing through the first, gives to the whole surface of the egg a wrinkled appearance, though the first shell is in reality entirely smooth. The large cavity, which is formed by these two outside shells, contains a fluid (which has an albuminous appearance, and turns milk-white on contact with water) in which swims the small globular embryo (measuring only 0·024 mil.) enclosed in a third shell, closely attached to it, but of considerable thickness (0·001 mil.). We cannot state with certainty that there are three pairs of spines to this embryo: if there are any, they must be very small."

The embryo of this species is as yet unknown; but Dr. Weinland suggests the possibility of the child having swallowed a fly, in which the embryos of this entozoon were contained; basing his suggestion upon the discovery of Stein, that in the stomach of the meal-worm beetle (larva of Tenebrio molitor) were found embryos of a tapeworm, which there afterwards became scolices, waiting until its host was eaten by another animal, which might perchance prove a suitable soil for its further development.

6. *Tænia capiensis.* — Under this name, Küchenmeister describes an entozoon, portions of which he obtained from Dr. Rose, of the Cape of Good Hope. These segments were without head or neck. "Its total length must be at least 6 to 10 yards. Its segments are very thick, white, and fat; in the mature state, more than 1 inch in length, 3 to 5''' in breadth, and extremely massive. They are distinguished by having a longitudinal ridge running along the whole of the mature and immature segments.

"The genital pores are irregularly alternate; the penis so much concealed behind the thick inflated margin of the genital pore, that it is hardly discoverable. The uterus is formed by a thick, median stem, into which 40 to 60 lateral branches open: these resemble those of Tænia mediocanellata, or perhaps still more those of T. ex cysticerco tenuicolli, especially when we consider the arrangement of the branches, like the teeth of a rake, at the upper and lower margins of the segments.

"The ova are oval, rather rounded, uneven, and 0·030 to 0·034 mil. in breadth by 0·038 to 0·040 mil. in length. They allow the six-hooked embryo, which is 0·024 mil. in diameter, to shine through them distinctly. I never saw such remarkably developed embryonal hooklets in any other human Tænia: the central ones resemble stilettoes. The inner hooklets were 0·0069 to 0·0071 mil. in length; the outer ones, 0.0046 mil. The calcareous corpuscles were as large and numerous as in T. mediocanellata.

"This Tænia is particularly rich in cholesterine; for very large flakes of this substance made their appearance in the deposit which I obtained from the sediment at the bottom of the bottle in which this tapeworm came from the Cape."

Of the migrations of the embryo and scolices, nothing is known; but he states, on the authority of Dr. Rose, that it is impossible that they should exist in the flesh of pigs, as Hottentots (from one of whom the specimen was obtained) abstain from eating pork as religiously as do the Jews and Mahommedans; that a thick Tænia also exists among the Mahommedan inhabitants in Abyssinia; and, more, that "it is known at the Cape of Good Hope, that the Hottentots brought this tapeworm with them from the Caffre wars, in which they enjoyed themselves amongst the cattle of the Caffres."

7. *Tænia echinococcus*. — This entozoon, of which comparatively little is known, is found quite often in Iceland, more particularly in its encysted state. It is small, seldom exceeding three mil., and possessing a circlet of hooks, 28 to 36 in number. It usually has but three or four articulations, the last of which is the mature segment, fecundated and charged with eggs. The sexual organs are lateral, the penis being found a little below the centre of the segment. The ovaries

are large and sinuous, and charged with spherically-shaped eggs. One quite noticeable feature in this parasite is, that the articulations, after separation, become larger than the entire Tænia (Moquin-Tandon).

This parasite, in the adult state, is found in large numbers in the dogs of Iceland; but whether in man, except as the "hydatid," is by some authorities considered doubtful.

8. *Tænia cucumerina* (T. elliptica, T. canina) is common to dogs and cats, and has been stated to have been found in the human body (Eschricht, Leuckart), although Moquin-Tandon considers this as doubtful. The scolex of this form is unknown. Dr. Weinland suggests that it may be found in flies, which dogs are so eager to catch and swallow.

9. *Tænia lophosoma.* — For a description of this entozoon, we are indebted to Dr. Cobbold, of London ("On Tapeworms," London, 1866), who found a specimen in the Museum of the Middlesex Hospital: —

"When complete, it must have measured 8 feet in length. It is characterized by the presence of a ridge, extending through the entire length of the body, imparting to most of the segments a pentagonal figure, when viewed from the front. The individual segments are much smaller than those of the full-grown tapeworms; and they are further characterized, collectively, by the presence of uniserially disposed, reproductive papillæ, extending along the left margin throughout the entire chain. The papillæ are very prominent, being placed at the centre of each joint.

The ordinary segments give an average breadth of $\frac{1}{5}$ of an inch, those at the caudal end stretching to as much as $\frac{3}{4}$ of an inch in length. Their greatest thickness does not exceed the $\frac{1}{8}$ of an inch. The eggs resemble those of other tapeworms, and offer a diameter of about $\frac{1}{850}$ of an inch."

Dr. Cobbold considers this worm as a new variety, differing from T. solium and T. mediocanellata, in the size of the segments and the uniform disposition of the papillæ; from Weinland's T. flavopunctata, in its large size; and from Küchenmeister's T. capiensis, in having the genital pores unilaterally situated. In his description, I should judge the head was wanting, as was Küchenmeister's; for at the beginning he says, "When complete, it must have measured," &c. Moreover, he does not describe the head at all. With the exception of the position of the reproductive papillæ, it agrees pretty nearly with Küchenmeister's description of the T. capiensis. Would that exception be sufficient to base a new genus upon? I would suggest whether it might not be identical with Küchenmeister's T. capiensis, especially if, as Dr. Cobbold remarks, it may come from mutton? Küchenmeister's specimen was supposed to have come indirectly from the cattle of the Caffres; and, of these, sheep form a large proportion. It certainly is an interesting question.

Of the Bothriocephali, there are now known to be two species: the B. latus (already described), and the B. cordatus (Pl. III. fig. 5), a new species, recently described by Leuckart, who received about twenty specimens from Godhaven, North Greenland. Of these, but one came from the human intestine; the remainder being from dogs, in which animals the parasite exists quite abundantly.

It is a foot in length; its head (Pl. III. fig. 5, A) is heart-shaped or obcordate, short and broad, and sits upon the body, without the intervention of a long neck. The segments (Pl. III. 5, B, a) are distinct from the very commencement near the head; and so rapidly do they increase in width, that the anterior part of the body becomes lancet-shaped. About fifty joints are immature; Leuckart, in the largest specimen, counted a total of 66 joints. The calcareous corpuscles of the skin are more numerous than in the B. latus (Aitken). The genital openings are, as in the B. latus, serially arranged in the centre of each joint; but the uterine lateral processes are more numerous (Cobbold, Aitken).

This completes the description of the adult forms of these parasites found in the human body; that of the Cysticerci now follows.

For the description of the Cysticercus cellulosæ, I refer the reader to the portion on the anatomy of the Cysticerci.

The *Cysticercus Tæniæ mediocanellatæ* (Pl. I. fig. 8) was for some time unknown; the question having been raised, as to whether it might not be, in the adult form, a T. solium which had lost its hooks from old age, until the experiments of Leuckart and Mosler in Germany, and Simonds and Cobbold in England, by feeding animals with the sexually mature segments of the T. mediocanellata, from which they obtained scolices without hooks, precisely resembling the head of the parent form, decided the question in the negative. As I have not the detailed experiments of the German observers at hand, I must refer to those of Dr. Cobbold, which I give from his work on Tapeworms:—

"It may interest the reader to explain briefly the nature and circumstances attending one of our experiments. In the case of the calf, I procured a quantity of the ripe or sexually mature segments of the unarmed tapeworm. These were immersed in warm milk, and introduced by the mouth. Sixteen days after the worm-feeding, some symptoms of infection showed themselves; but, in a few days more, they entirely subsided. A second administration of the worm segments was therefore decided on. Again, fifteen days after the second feeding, fresh symptoms of irritability supervened; and, for a few days, the distress of the animal seemed to forebode the likelihood of a fatal result. However, after a while, its condition improved: the general expression of the face indicated returning health, the breathing and pulse improved, the tremors subsided, and the appetite returned. Convalescence being perfectly established, the animal quickly gained flesh, and, in two months' time, might have been sold to a butcher as a perfectly healthy and well-nourished animal. In truth, it was healthy; only, as we shall presently see, its body was full of parasites, resulting from the worm-feeding. About three months after the date of the first administration, the calf was slaughtered. The flesh was carefully examined, and, according to my estimate, contained no less than 8,000 measles. These measles were undoubtedly the young of the unarmed tapeworm, presenting as they did the essential characteristics which I have already described; viz., the absence of the crown of hooks; the broad, flat head and neck; and the absence of the prominent rostellum."

In looking over the first volume of Aitken's "Science and Practice of Medicine," I find a *resumé* of the experiments of Professor Leuckart of Giessen (whose results are the same as those of Dr. Cobbold), and which I transcribe as proof additional of this point:—

"He fed two calves with the fresh eggs of the T. mediocanellata, by giving them the proglottides of this parasite. The first animal he experimented on died from a violent attack of the measles disease; and, on dissection, the muscles were found filled with measles or vesicles, containing imperfectly developed scolices. On the second day, a smaller number of proglottides (in all about fifty) were administered, and the febrile symptoms again appeared with such violence, that Leuckart thought this animal would die also. Fortunately, after the lapse of a fortnight from the commencement of symptoms, some abatement of the disease took place; and this gradually continued until the animal was perfectly restored to health. Eight and forty days subsequent to the earliest feeding experiments (which were continued at intervals for eighteen days), Professor Leuckart extirpated the left cleido-mastoid muscle of the calf; and, while performing the operation, he had the satisfaction of seeing the cysticercus vesicles lodged within the muscles. They were larger and more opalescent than those of the Cysticercus Tæniæ cellulosæ, but nevertheless permitted the recognition of the young worms through their semi-transparent coverings. The heads of the contained Cysticerci exhibited all the distinctive peculiarities presented by the head of the adult strobila, the T. mediocanellata."

From these experiments there can be no doubt as to the fact, that the T. mediocanellata is not an old T. solium, nor as to whence it is derived. Besides the appearance of the head and neck, the distinctive mark between the scolices of this Tænia and those of the T. solium consists in the larger size of the vesicles.

3. *Cysticercus tenuicollis.* This scolex is considered as the embryo of the T. marginata, found in the intestines of dogs, and probably of wolves. As, in the adult state, it is doubtful if it exists in the intestine of man, I have omitted the description of the Tænia.

The head resembles somewhat that of T. solium, except in the form, number, and size of the hooks, which are more slender, less strongly curved, and remarkable for the extraordinary length of the stem. They are arranged in two series, from 32 to 42 in number, and, according to Leuckart, measure 0·178 and 0·114 mil. in length, while Küchenmeister gives 0·175 to ·215 and 0·117 to ·126 mil. The caudal vesicle is large (1, 2, 4, 6", and more), frequently enormous (sometimes attaining, in animals, the size of a child's head), of extreme thinness, and is remarkable for the concentric wrinkles visible externally, and which are crossed by fine longitudinal striæ, giving it a chequered appearance (Küchenmeister). The body is from 14 to 30 mil. and more in length, and from 5 to 10 mil. in breadth, broad and cylindrical; the neck is from 8 to 15 mil. long.

Even in the form of the scolex, this parasite is rare; so that, by some modern authors, it has either been stricken out of the list, or else considered as a doubtful Helminth. Eschricht, of Denmark, quotes in all five cases of the occurrence of this parasite in the human body; one reported by Kolpin, one by Treutler, one by Zeder, and two by Schleissner. They occurred in Iceland. Küchenmeister, however, contends that but one of the cases reported (Schleissner's) was that of true C. tenuicollis, while the others were the young of other species. " This is a mistake which can easily be made, inasmuch as, when the enveloping cyst is unopened, there is sometimes hardly any external difference between the Echinococcus and the C. tenuicollis, especially when the latter is in the liver."

3. *Cysticercus acanthotrias* (Pl. III. fig. 6). For the description of this parasite we are indebted to Dr. Weinland, who found it in the collection of the Society of Medical Improvement, of Boston, Mass. It was supposed to have been a cysticercus cellulosæ, and was obtained from the body of a white woman, æt. 50,

who died of phthisis in Richmond, Va. "About twelve or fifteen cysts were found in the cellular membrane of the muscles and in the integuments, besides one which hung free from the inner surface of the dura mater, near the crista galli. In the same subject there were also numerous specimens of Trichina spiralis." — *From Dr. Jeffries Wyman.*

Upon examination, Dr. Weinland found, that, instead of two rows of hooks, the usual number, there were three. He therefore proposed for it the name of Acanthotrias (ἄκανθα, τρεῖς).

His description I copy from his "Essay on Tapeworms:" —

"There are fourteen hooks to each of the three rows, and the hooks of the different rows are different in shape and size. The row of the largest hooks (Pl. III. fig. 6, B) is the innermost, when the muzzle is stretched out. These resemble the large ones of the true Tæniæ: their total length is 0·153 mil. The length of their foot, as I call that part which, being subservient to a firm hold, is buried in the tissue of the muzzle, is 0·090 mil. The hooks of the next or second row (Pl. III. fig. 6, C), standing between those of the first row, and projecting with their thorns (that is, the free parts of the hooks) a little beyond the thorns of the former, are a good deal smaller, but of the same general shape: their total length is 0·114 mil., and the length of the foot 0·063 mil. The hooks of the third row (Pl. III. fig. 6, D), lying outwards, yet between the two former rows, are far the smallest, their thorns more curved, and their feet much shorter:

their length is 0·063 mil., and the length of the foot only 0·030 mil.

"The calcareous corpuscles in the skin of the worm (Pl. III. fig. 6, A) measure 0·003 to 0·009 mil. The length of the whole Cysticercus is about 10 mil., without the bladder, which latter is of about the same size as in the C. cellulosæ. The suckers of the head are visible to the naked eye. The neck is irregularly scalloped close to the head. At a distance of 5 mil. from the head, the worm grows suddenly broader; and there are distinct folds running transversely over the worm, dividing this into distinct joints. To this broader part is attached the bladder, by a narrow bridge. The cysts, which contain Cysticerci, have exactly the same appearance as those of C. cellulosæ; looking somewhat like white beans, they lie along the fibres in the muscles, splitting them in such a manner as to make a gap before and behind the cyst, which is filled with yellow fat."

Thus far, I have given the list of the Tæniæ and Cysticerci now known to inhabit the human body. Those in the list left for me to speak of are the Echinococci and Acephalocysts.

Of the former group, there are two kinds,— the Echinococcus hominis (E. altricipariens of Küchenmeister), and the Echinococcus veterinorum (E. scolici pariens of Küchenmeister). Of these two, the latter is the only one of which the mature form is found, viz., the Tænia Echinococcus (Von Siebold); that of the former (according to Küchenmeister) being unknown. As the Tænia is an inhabitant of the intestines of dogs, and not of the human subject, I shall not dwell upon a description of it.

The scolex is contained in a vesicle which hardly exceeds the size of an apple, and, according to Eschricht, generally measures $2\frac{1}{2}$ to 3 inches. This vesicle consists of two similar membranes, arranged in several concentric lamellæ, of which one (the outer) is thick and cartilaginous; the inner one, thin and membranous. From the inner surface of this latter are seen small elevations, up to $4'''$, which are partly very young scolices in the act of development, partly further developed scolices of $10'''$, and partly representing the points where such scolices formerly sat; sometimes contained in cysts, sometimes not (Eschricht, Küchenmeister).

Within the fluid contained within this vesicle are seen free scolices (resulting from the rupture of the cysts, or their detachment from the membrane), some furnished with the stalk by which they were attached, and, on being pressed, form flat discs. From the hinder portion near the stalk is seen a circlet of 30 to 40 hooks, arranged in a double row; at its anterior extremity, four sucking discs; and within these latter the calcareous corpuscles, measuring 0·01 to 0·02 mil. (Küchenmeister).

2. *Echinococcus hominis* (E. altricipariens of Küchenmeister). — The mature form of this scolex is as yet unknown, although Küchenmeister thinks that it may probably exist in the human intestine, and has suggested that the T. nana of Von Siebold has such an origin, its scolex being nothing but

an E. altricipariens. In this species, although the enveloping cyst and the mode of annexation of its inner wall to the first vesicle of the Echinococcus are the same, it is distinguished by its greater size. Again, in this form we have not only single scolices or single vesicles, but the production of secondary and tertiary ones (daughter and grand-daughter vesicles), sometimes with, and sometimes without, the production of separate scolices, adhering directly to the walls of the vesicles (Küchenmeister). The scolices thus produced (the mother, daughter, or grand-daughter vesicles) are usually more slender; show marked sucking discs; have a larger number of hooks (45 to 54 in number), which are more slender than those of E. scolicipariens, and arranged in a double row.

This scolex is found in various portions of the human body, — in the liver, lungs, kidneys, scrotum, sheath of the testis, the spleen, ovaries, breast, throat, in the subcutaneous cellular tissue, and in the bones. It may attain a large size (that of a child's head), and is the cause of the Icelandic disease, the so-called Echinococcus disease. As an estimate of the frequency of its occurrence, Dr. Schleissner found in some parts of the island, on an average, two or three members of each family infected; and Dr. Thornstein says that every seventh man in Iceland has this disease (Weinland).

The last group under consideration is that of the Acephalocysts; and, as these are now known to be degenerate or imperfectly developed Echinococci or Cysticerci, we might presume, *a priori*, that the different kinds are but barren cysts of either one or the other species. Such they are; and Küchenmeister in his work refers them as derived from —

1°. Tænia Echinococcus scolicipariens.
2°. Tænia Echinococcus altricipariens.
3°. Tænia ex Cysticerco tenuicolli.

To the first class he refers those acephalocysts which bear no daughter vesicles in their interior.

To the second class he refers those acephalocysts which have a formation of daughter vesicles, but no scolices; and he observes, that "the only remarkable thing, perhaps, is the circumstance, that the enveloping cysts of acephalocysts, with clear watery contents and of small size, are thicker and more cartilaginous than those of the true proliferant Echinococci, in which a similar structure of the cyst usually occurs only in large colonies, or in those which contain the remains of dead scolices and purulent grumous masses. And,

In the third case, Küchenmeister, while administering to a lamb the eggs of a T. ex Cysticerco tenuicolli, found a sterile C. tenuicollis in the midst of hundreds of other equal sized and fully developed Cysticerci of this species. This sterile one bore perfectly distinct indications of life; and he furthermore says, that, although he knows of no other specimens of Cysticerci or Cænuri having been met with in a living but barren state, yet it undoubtedly is a possible thing.

PART V.

Development of the Embryo into the Tænia.

HAVING thus far described the anatomical characters of the Tæniæ and their embryos in their different stages of development towards the perfect state, as also the different species of the human cestoidea, it would not be amiss, before leaving the subject, to follow the different steps of their transition towards the perfect state.

It has already been shown, that the eggs are contained in the proglottides, and that these are thrown off in the fæces. The ova are extruded from the ripe proglottides, even while in the intestine of the bearer of the tapeworm; but here they are never hatched, for the reason that they do not find provisions suitable for their development. These ova, from the tough, resistant nature of their shells, we have seen, are admirably calculated to withstand injurious influences, even after their containing proglottis has died and undergone decomposition. They find their way into the intestine of their future host, usually through the medium of food and drink: the shell is ruptured, partly by the act of mastication, and partly by the solvent action of the gastric juice, and the embryo is set free. As before mentioned, the embryo is furnished with six hooks, arranged in three pairs, — one in front and one at each side (Pl. II. fig. 1, a). Those in front bring themselves together into a point, and move backwards and forwards; the two lateral ones are armed at the apex with a pair of hooks, and move sideways. According to Van Beneden, progression is accomplished by bringing the two centre hooks into a wedge-like shape, thrusting and twisting, while at the same time it assists itself with the two lateral pairs of hooks, and with them pushes itself onwards, "just as a person who wishes to spring out of a low window rests his elbows against the window-frames, and drives himself forward with a spring;" but Leuckart thinks that the movements are by no means effected right and left in all the hooks in the same direction, but that, " whilst the two lateral pairs move downwards from the vertex in the lateral plane nearly simultaneously, this movement can only take place somewhat later in the central pair, and then in the median plane." Thus sometimes one pair of hooks (the central one) appears to rest, and to hold the embryo in its place, whilst the two lateral pairs are in motion; and sometimes the central pair appears to be active, whilst the other pairs hold the embryo and preserve it from slipping back, which certainly appears most in accordance with nature (Küchenmeister). Whichever the method of boring be, the embryo quickly penetrates through the walls of the intestine into the tissues beneath, where it takes up its abode. This may be done entirely by the active method of migration, viz., boring, more particularly in those embryos which are developed in cold-blooded animals; but, in those which are found in the warm-blooded animals, a passive as well as an active migration takes place. The embryos bore through the tissues into the blood-vessels, and are then carried along by the circulation, until they bring up in some of the smaller vessels, which they then leave as they entered. Leuckart four times found free embryos in the blood of the main branch of the vena porta at its entrance into the liver; and Küchenmeister also, in speaking of this subject, remarks, that he is " convinced that the brood is introduced into the circulation of the blood by the active penetration of the tissues, and then a passive migration with the blood is commenced." The supposition, entertained by Küchenmeister and Haubner, that these embryos penetrate the abdominal cavity and the organs therein contained through the ductus choledochus, has not been supported by Leuckart; and there is reason to believe it is unfounded.

To sum up the manner of the migration of these embryos: —

1°. A portion of the six-hooked brood in all species of Cestoidea (whether in cold or warm-blooded animals) may reach their dwelling-place directly, and by active migration.

2°. Another portion, after a longer or shorter migration, reaches the vascular system of the new host (the blood-vessels, and perhaps also the lymphatic system, as, according to Virchow, is the case with the Echinococci); is subjected to a passive migration here with the fluid; remains fixed in the smallest ramifications; and, —

α. Becomes further developed in the vascular tube, making use of the walls of the vessel as an envelope (cyst), so as to remain there permanently; or, —

β. Migrates passively into the neighboring tissues, after the rupture of the walls of the vessels, in consequence of the swelling of the body of the embryo; or, —

γ. After sticking in the finest vascular ramifications, enters anew upon an active migration by means of its six small hooklets, passing through the walls of the vessels into the neighboring tissues, and often continuing its wanderings for some time in closed cavities, or in the soft parenchyma of organs, even when it has attained to a certain size, so as to be visible to the naked eye (Küchenmeister).

Having reached, by these wanderings, the suitable spot for its further development, we then notice that the embryo becomes stationary; and frequently we are able to track it by streaks of a yellowish color, which are produced in the organ by exudative inflammation, caused by these journeyings. This exudation sometimes also occurs round the embryo itself, enclosing it in a capsule, and excluding it from the organism: this process has been described by Von Siebold ("Ueber die Band-und Blasenwürmer"), under the name of the "extrinsic encysting process." When this extrinsic encysting process does not take place, the embryo begins to enlarge and swell up from the fluid nutriment which it receives from the tissues where it is imbedded. When it has attained to a certain size from imbibition (supposing it to be in the interior of a solid organ, and not in any serous cavity), it surrounds itself with a capsule or cyst to protect it from external pressure, and to allow of its obtaining the repose requisite for its further development. These cysts constitute an absolutely new formation, which commences round the germ of the vesicular worm, from the same masses of exudation which are deposited around the youngest brood during its migration. This new formation acquires a structure analogous to that of the subjacent tissues, and constantly increases in size with the growth of the young cestode vesicle.

When, however, the young arrive in free cavities, this enveloping cyst is either absent, or, if present, is not developed until a later period, and at a time when the embryo has become a vesicle of considerable size (Küchenmeister).

While these embryos are increasing in size from the cause above mentioned, it is seen, that, at the anterior end, and near the point where the six hooks are situated, a depression occurs, a funnel-shaped pit is there formed, and this gradually penetrates deeply into the body of the embryo. In this pit, the head first makes its appearance, its lateral walls forming the body of the future worm; and the rest of the vesicle, which is not inverted, becomes the so-called caudal vesicle. That these embryos are really developed into the scolices, is shown by the discovery of Stein, of the embryonal hooks upon the embryos themselves while increasing in size, and also upon the cestoid vesicle, when developed into the vesicular stage; and it also shows that these hooklets do not enter into the formation of the scolex.

When the head begins to form, the caudal vesicle ceases its activity in the Cysticerci: it then becomes only the nutritive receptacle and the protective organ for the head.

At this point its development ceases, at least in the tissue where it has become a scolex. In order for it to progress to the adult state, it must pass into the *intestine* of some higher animal. This is usually accomplished by the bearer of the scolex falling a prey to the larger animal; and the scolex is thus passively introduced into the intestine of its new host. Should this, however, not happen, the scolex advances no farther, but remains a scolex.

Being introduced into the stomach of the new host, the tissue of the animal in which the scolex is contained undergoes digestion and dissolution; and the scolex, with the vesicle, is set free. The enclosing vesicle is either dissolved during the process of digestion, or else is opened by mastication; and then the cystic worm escapes into the cavity of the stomach. The caudal vesicle now collapses, either from exosmosis, or by the solvent action of the gastric juice or other digestive fluids; and, in about five or six hours after the ingestion of the worm, it has found its way into the small intestine. The head and neck then extend themselves by a species of evolution, by which process the margins of the head and neck, which had previously been inverted, now become the free borders of the worm. The partially digested body and caudal vesicle now become detached from the head and neck; and it is often seen, within the first twenty-four hours, that the adhering scolex bears behind it the middle body and the caudal vesicle on a fine filament, *i.e.*, a rudimentary transverse fold or so-called segment (Küchenmeister). All that is left of the scolex, therefore, is the head and neck, and this filament. This latter, in process of time (about two days; Küchenmeister), likewise becomes detached, and its site is marked by a cicatricial notch. This notch, according to Leuckart, leads into a cylindrical cavity, passing through the whole body as far as the rostellum; its walls grow greatly, by which means the scolex is converted into a solid and not inflated body. After a varying period, the segments begin to be formed, and the worm has then arrived at its adult state. (For the time at which the sexual organs are formed, and at which the Tænia becomes perfect, I refer the reader to the first portion of this essay.)

In the preceding, it has been supposed that the scolex had arrived at an intestine favorable for its development. Should the opposite be the case, the process goes on as far as the segmentation: beyond that point, all the growth consists in an inarticulate tail-like appendage (Küchenmeister).

The formation of the proglottides or segments is always from the neck downwards, by a budding forth from the body, without sexual propagation; and this becomes constricted into segments by transverse furrows or wrinkles. It therefore follows, that the older or more matured joints are the farthest from the head; and the cicatrized notch is always visible in the last joint during the process of growth, until this latter is cast off. Its absence, therefore, indicates that segments have been cast. The one bearing the notch, however, is generally sterile (Küchenmeister); but the others are not thrown off until the joints are ripe and fecundated.

Before leaving this subject, it will be well to say a few words about the embryo of the Bothriocephalus (Pl. II. fig. 3, A). The scolex is not known; neither is the process of its development. The embryo is ciliated, lives in the water, and is supposed to be developed in certain kinds of fishes; as, for instance, the so-called Bothriocephalus solidus, which is found in the abdominal cavity of the stickleback. This fish is eaten by some larger fish or by some waterfowl, and the scolex is then further developed into the mature worm, *e.g.*, the B. nodosus, which Von Siebold states to be nothing but a further development of the B. solidus of the stickleback. It is therefore probable, that the same condition of things obtains in others of the Bothriocephali.

We have now followed the embryo in all its wanderings, — from its exit from the ovum, to its *début* as an adult tapeworm. Interesting as it would be to discuss some of the experiments performed by various observers in confirmation of the statements made with regard to the development of the Tæniæ from the Cysticerci, my time and space will not permit me to here do further than to allude to them, by referring those interested in such matters to Küchenmeister's work, "Die in und an dem Körper des lebenden Menschen vorkommenden Parasiten;" Von Siebold's "Ueber die Band-und Blasenwürmer," and those of others, for a detailed account of their experiments with the young of the various species of the Tæniæ. They gave the scolices, and developed with them the perfect worm; thereby establishing the connecting link which binds together such apparently different forms.

PART VI.

Pathology and Treatment.

ATTEMPTING to classify the symptoms which arise from the presence of these tapeworms, is somewhat like classifying those arising from the so-called disease, hysteria; and, in this connection, permit me to suggest, that many cases of hysteria may possibly arise from the presence of this parasite, more particularly in women. Their name is legion; and withal so obscure are they, and so calculated in their form to draw the medical man's attention away from the true cause, that, until the detection of the proglottidès in the fæces, the diagnosis must, in a great measure, be conjectural.

Often, these entozoa are not productive of any marked trouble to their host; and indeed in Abyssinia the often-quoted remarks, that no person is considered healthy who is not the bearer of at least one tapeworm, and that no slave is sold without a packet of Kousso, would tend to make it appear, that, in that country at least, no evil results from their presence. Still, it oftener happens that they are not so innocent; and the feeling of weariness and lassitude which often occurs may be attributed to them, not to mention the restlessness, nervous irritability, and headache, which is often attendant on the first train of symptoms.

In more severe cases, the headache increases, and is often accompanied by vertigo, noises in the ears, pruritus about the anus, mouth, and nostrils, inducing constant scratching, obscure pains about the body, loss of appetite, sleep, and dyspepsia. As a natural sequence to this latter, hypochondriasis results. Dr. Cobbold, of London, speaks of one peculiar symptom in this connection, viz., a tendency to faintness, which "is so marked as to create much alarm; and a person, uninformed of the true cause of the disease, might be led to treat the symptoms as arising from a totally different source." Indeed, I cannot conceive of a symptom so likely to withdraw the medical attendant's attention from the true cause of the trouble, as this one symptom; referable as it is, more especially in women, to so many diseases of a functional, as well as of an organic, nature.

The bowels, as might be expected, are frequently the seat of much pain, sometimes spasmodic and gnawing, but more generally dull and heavy; akin to the feeling of weight observed in dyspepsia. Diarrhœa is at times present, though it is by no means constant, the opposite effect sometimes being produced; the stools, exceedingly dark, or, it may be, entirely lacking in color. It often happens also, as a result of so many combining, irritating causes, that a remittent fever is induced, which is known under the name of "worm fever" (Aitken).

Truly, it may be considered that these are severe

troubles to arise from apparently so slight a cause; but the worst yet remains to be told. In the worst cases, severe functional disturbances of the nervous centres manifest themselves, more especially connected with the brain and spinal cord, and showing themselves in the occurrence of epilepsy, convulsions, and chorea. This latter symptom, together with hysteria, is more especially noticed in the female sex. Instances are not wanting in which mania has been attributable to the presence of Tæniæ in the intestinal canal (Cobbold). Neither have the nerves of special sense escaped from harm; and cases have been reported where these worms appear to have caused strabismus and amaurosis. Dr. Shrady reported to the East-River Medical Association, of New York, a case of blindness from the same cause, which was cured by the expulsion of the worm ("N.-Y. Medical Record," vol. i. p. 483).

The embryos of these parasites, whether in the form of Cysticercus, Echinococcus, or Acephalocyst, may also be the cause of trouble; but more especially when situated in the cerebral substance, where, by their presence, they give rise to convulsions, epileptiform seizures, and the like, as well as to injury of the visual organs, when they are situated in that portion of the nervous system. In the liver, they give rise to icterus (from pressure upon the biliary ducts), to ascites, œdema of the extremities, anasarca, &c. (from the obstructions they cause to the portal circulation). In addition to this, the passage of the embryos may give rise to fever, vomiting, and even death.

Treatment.

Over the Tæniæ we fortunately have control (when their presence has been diagnosticated); over their embryos we have none, even if we should be able to suspect their presence. Numerous have been the remedies proposed for this parasite. Of these, I shall select the best for discussion, as it would not be to the purpose of this paper to enlarge upon *all* the known curative agents, many of them uncertain in their action.

Of these, Oil of turpentine (Ol. terebinthinæ), Kousso (the flowers and unripe fruit of Brayera anthelmintica), Kamela (K. wurrus, the powder adhering to the capsules of the Rottlera tinctoria), Filix, (the dried rhizoma of Aspidium filix mas), Granati radicis cortex (the bark of the root of Punica granatum), and Pumpkin seed (the seed of Cucurbita pepo), are among the best and most highly esteemed as anthelmintics.

Turpentine (Ol. terebinthinæ). — This remedy has been spoken of in the highest terms as an anthelmintic, on account of its applicability to nearly, if not quite all, the varieties of tapeworms; its rapid action; and because it expels the Tænia entire and in one piece; which latter quality Küchenmeister regards as " a requisite of a good vermifuge." As an offset to its good qualities, it is liable, if given in too small doses, to produce sickness, inclination to vomit, ulceration of the mouth, griping pains, and suppression of urine; and, if in too large quantity, bilious stools, tenesmus, and bloody stools and urine (Küchenmeister). Its action may be facilitated by the exhibition of Ol. Ricini; and should the symptoms above enumerated supervene, the administration of the Turpentine must be suspended, and that of Castor oil commenced (Copeland). Thom. Schmidt, in Clarus' " Arzneimittellehre," p. 703, says it should never be given in large doses in winter and in moist, cold weather, as, under such circumstances, it has not a purgative, but a heating, action (Küchenmeister). The methods of administration most recommended are : At bedtime, pure in the dose of ʒj ; or else triturated with ʒj Castor oil, or with one or two drops of Croton oil, or two or three yolks of eggs with ʒj purified honey, given in two or three portions in the course of 1 to 1½ hours. For children, but half the quantity is requisite (Küchenmeister).

Kousso (flowers and unripe fruit of Brayera anthelmintica).—This is well known as the Abyssinian remedy, and has been highly extolled for its virtues. This, however, has not stood the test of experience as well as was first expected; probably owing, partly to its frequent adulteration, and partly to the drug being kept too long. This latter point is one often overlooked, and is moreover a very important one. Most remedies, and especially those dependent upon some volatile principle for their action, should never be used except in a fresh condition. Küchenmeister's experiments, show that, in a decoction of Kousso flowers mixed with albumen of egg, the Tænia crassicollis of the cat died in a short time. This drug, it must be remembered, even when given in the ordinary manner, is liable to produce sickness, violent pain in the intestines, and vomiting.

The dose of the powder is from ʒvj to ʒj, usually given in infusion.

The methods of Professors Martius of Erlangen, and Raimann of Vienna, are well spoken of by Küchenmeister. The former, finding that the powder killed the worm, but that it did not cause the head to pass away, endeavored to isolate the active principles of the resin. "A red resin obtained from Kousso had no effect. It was otherwise with a soft resin of the Kousso, of which ℈ij were obtained from ʒvj of Kousso, but in which there was certainly still some red resin and a waxy substance. This soft resin, or more correctly resinous mixture, was dissolved in alcohol at 36° R. (113° F.), and filtered; the alcoholic solution was dropped upon sugar. As soon as the alcohol was evaporated, the solution was again poured upon sugar; the whole was well dried, and reduced with sugar to the finest powder; sugar being added until, with ℈ij of soft resin, the whole quantity weighed ʒss. This very finely divided resin was mixed with ʒj of honey, and the whole administered in a period of 12 to 16 hours, commencing at four in the afternoon. The next morning an aperient was given, Castor oil or a salt."

Küchenmeister, in using it in this way, found in three cases that the worm was expelled up to the neck, but in such a detached manner as to render it impossible to find the head. In one of these cases, segments of the worm were passed at the end of three months. Still, although he admits the efficacy of this method of administering the drug, from the fragmentary state in which the worm is passed he cannot give it the preference over Turpentine or Pomegranate root.

The method of Professor Raimann, of Vienna, is as follows : " ʒvj of Kousso are macerated for 24 hours, and then boiled for half an hour. This infuso-decoction is taken while fasting in two portions, without straining, and therefore the flowers in it; and, two hours afterwards, ʒj to ʒij of Castor oil" (Küchenmeister).

From reports in Hebra's "Zeitschrift," the remedy is well borne, and acts with certainty.

Kamela (K. wurrus or Reroo, the powder adhering to the capsules of Rottlera tinctoria) is another comparatively new remedy, and has received special notice from Dr. Alexander Gordon, Surgeon of the Tenth Regiment of Foot, who had used it in India. He says : —

" The success and rapidity of effect of the Kamela in removing tapeworms in the cases of soldiers of the Tenth Regiment, to whom I administered it, were such that I did not consider it worth my while to keep notes of them after the first two or three. Nor, indeed, were the men to whom it was administered latterly taken into the hospital ; for they soon became aware of the wonderful efficacy of the remedy, asking of their own accord for a dose of it, after which they invariably parted with the worm in the course of a few hours, and then went on with their military duties as if nothing had happened; while, as I afterwards ascertained, considerable numbers did not think of 'troubling the doctor' at all, but, on suffering from the characteristic symptoms of the worm, applied for the Kamela to the apothecary, and always with the same effect. With Kamela, there is no unpleasant effect. It is not even necessary to take a dose of purging medicine as a preparative; and, beyond a trifling amount of nausea and griping in some instances, no unpleasant effects are experienced; while by far the greater number of persons to whom it is administered suffer no inconvenience whatever, beyond what they would from a dose of ordinary purging medicine.

"It was usually given as a spirituous tincture, by adding to ℨiv of the powder Oj of alcohol, and then filtering. In this manner ℨvi were obtained, and of this from ʒj to iij were given in a little mint-water. It seldom happened that more than ʒj was found requisite;" and Dr. Gordon says " he has never seen the remedy fail in removing the worm in a case where there were unequivocal symptoms of its presence." The worm was always discharged dead; but it was at times difficult to state with certainty whether the head was in all cases passed with the rest of the body of the animal, although the Doctor is almost positively certain, so far as the eye could judge, that it was. ("Medical Times and Gazette," May 2, 1857).

Male Feru (Root of Aspidium filix mas). — This root is more efficacious as a remedy for the Bothriocephali than for the Tæniæ. The best form to give it is in that of the Etherial Extract, from 30 to 50 drops, half to be taken at night, the rest in the morning; the patient having previously fasted to be followed in an hour by ℥jss of Castor oil (Mayor).

When given in the form of a powder, Blosfeld's and Rapp's method is perhaps as good as any. "On the previous evening, a thick paste of bread and milk. In the morning, ℨj Pulv: Rad: Filic: Maris: is given in ℥jss of nutmeg tea. After six or eight doses, the worm is expelled. The use of Panna (Aspidium athamanticum) does not possess any advantages over the A. filix mas, and it is doubtful if it be as efficacious. Küchenmeister says, in three cases in which he administered it or saw it given, it did not fulfil the expectations which might have been entertained with regard to it. In the first case, the patient received no benefit from two doses of it internally (besides its exhibition as clysters); proglottides still being thrown off at the end of nine months after. In the other two cases, fragments were brought away (in one of the cases at the end of twelve hours); but no head could be found.

The first two were cases of T. mediocanellata; the third, one of T. solium.

Besides this, it possesses the disadvantages of being slow in its action, requiring a tedious preliminary treatment, and producing sickness, vomiting, and congestion of the head.

Should it, however, be considered advisable to give this drug, he recommends that the following rules, after Dr. Behrend, should be followed:—

Preliminary Treatment.

"For three or four days before the cure, nothing but easily digested food is to be taken; all sorts of flour, gruel, and cakes are to be avoided, as well as potatoes, and all spirituous liquors; because to dogs, to which the latter were administered, the Panna produced no result.

"For habitual costiveness, Carlsbad salts or lavements are to be administered.

"*Day of Cure.*—In the morning, fasting, every quarter of an hour 20 to 30 grs. of Panna powder in as little water as possible or in light beer, until ʒj to ʒss have been consumed, according to the age and condition of the individual. If the bowels are not moved, Castor oil is given some time after the last dose. It is best given two hours after the last dose. If the oil be given too early, the worm is torn; if too late, it is also injurious; and if the bowels are moved too violently, or too much of the medicine be given, the worm passes away in fragments."

Küchenmeister rather ridicules its use as a remedy; and certainly, from the above rules, it would seem as though it were not an altogether reliable medicine.

Pomegranate root (Pun: Gran: Rad: Cort:). This remedy is prepared from the bark of the root, which is collected in spring, and dried in the shade. It is frequently adulterated, principally with the box shrub (Buxus sempervirens), the barberry (Berberis vulgaris), and the caper (Capparis spinosa). In this drug, it is absolutely requisite that the bark be fresh, as it acts more gently than the dried, although more is required, and, as by keeping, it loses its active properties.

"The most efficacious form of administration is the solution of the extract in a certain quantity of water. The extract itself, made into an electuary with honey, or administered in pills, is to be recommended where there is a great tendency to vomit; but, on the whole, its aqueous solution is better."

Küchenmeister's method of giving this drug is the one first practised by Von Klein, of Stuttgart; viz., in combination with Extr: Filic: Mar: Ether, in the following manner:—

" ℞. Extr: Rad: Punic: Granat: Aq: quant. adep. es ex rad. ℥iv–vj; solve in aq. distill. ferv., ℥vj–viij.— Adde: Extr,: Filic: Mar: Ether: ℨj–ʒss; Extr: tanacet: vulg: ʒij; Gi guttæ (Gambog.) gr. iv, vj, ad x. — Shake.

"S. — A cupful to be taken in the morning (6 or 7 o'clock), fasting; a similar dose in three quarters of an hour. The third is kept in reserve. If the worm should not be expelled in 1½ hours after the second dose, the last portion is also taken. If vomiting occur, a table-spoonful of the medicine is given every ten minutes. To alleviate the tendency to vomit, the patient should gargle after every dose with fresh milk, but without swallowing any of it. Between the doses, also, he may take as much Elæo-sacchari[1] citri as will lie on the point of a knife, as often as he likes. If no evacuation have taken place three hours after the first dose, and the worm has not been expelled, an aperient is administered. With Tænia solium, Castor oil is usually sufficient, one or two table-spoonfuls every half hour or hour, or Gi guttæ, gr. vj to viij. Pulv: jalap: gr. x–xv; to be taken at once."

---

[1] In the German Pharmacopœia, an Elæo-saccharum consists of Sacchari albi ʒj et Guttæ xxiv of the required oil (Weinland); *e.g.*, Elæo-sacchari citri =Sacchari albi ʒj; Ol: citri limon: gtt. xxiv.

"*Subsequent Treatment.* — None, except tonics in cases of great weakness.

"*Preliminary Treatment.* — At the season of fresh strawberries and grapes, I give Oss of the fresh fruits every morning, fasting, for six to eight days; and, on the evening before the expulsion, a herring salad, with plenty of vinegar, onions, raw and boiled ham, and plenty of oil, and to very costive persons ʒj of Castor oil; after which, the patient may drink a large glass of light Rhenish wine, or a glass of Bavarian bitter beer. If the fruits cannot be had, the salad alone must suffice" (Küchenmeister).

For very obstinate cases of T. mediocanellata (which he considers the most difficult ones to expel), Küchenmeister recommends "so much of the ordinary electuary lenitive of the English Pharmacopœia (Confectio Sennæ, U. S. Br.), with the addition of Extract. tanacet. vulg., ℨij to the ʒj of the electuary as will produce a couple of soft motions daily; and then to take the mixture, but not before. Fasting the night before is bad: the medicine does not agree well with an empty stomach.

"If, after this treatment, portions of the worms should depend from the anus, a cup of strong, black coffee should be given, and, if needful, an aperient of calomel and jalap."

Pumpkin Seed (the seed of Cucurbita pepo).— The method with this drug is one much used in this country, and has been well spoken of. The best method of administering it is as follows:—

"Procure sufficient of the seed of the pumpkin (those grown in the West Indies are the best) to make ʒij, after the removal of the outside shell of the seed; put them into a mortar, and add Oss of water; pound them well up, and make a liquid orgeat of them, which strain through a cloth. Drink this mixture in the morning, on a fasting stomach. If it does not operate in the course of an hour and a half, take ʒj castor oil. Drink all the time as much fresh cool water as the stomach can bear or contain; that is, drench yourself with water. After taking the orgeat, if the stomach be well rubbed with ether, and an injection of about gtt. Lx of it be taken, it will be found to assist the orgeat; but this may not be necessary. Should the first application of the remedy not answer, repeat it the next morning, and there is no doubt your complaint will be removed. The worm will leave the patient at once, and probably entire" (Boston Medical and Surgical Journal, Oct. 3, 1851).

PLATE 1.

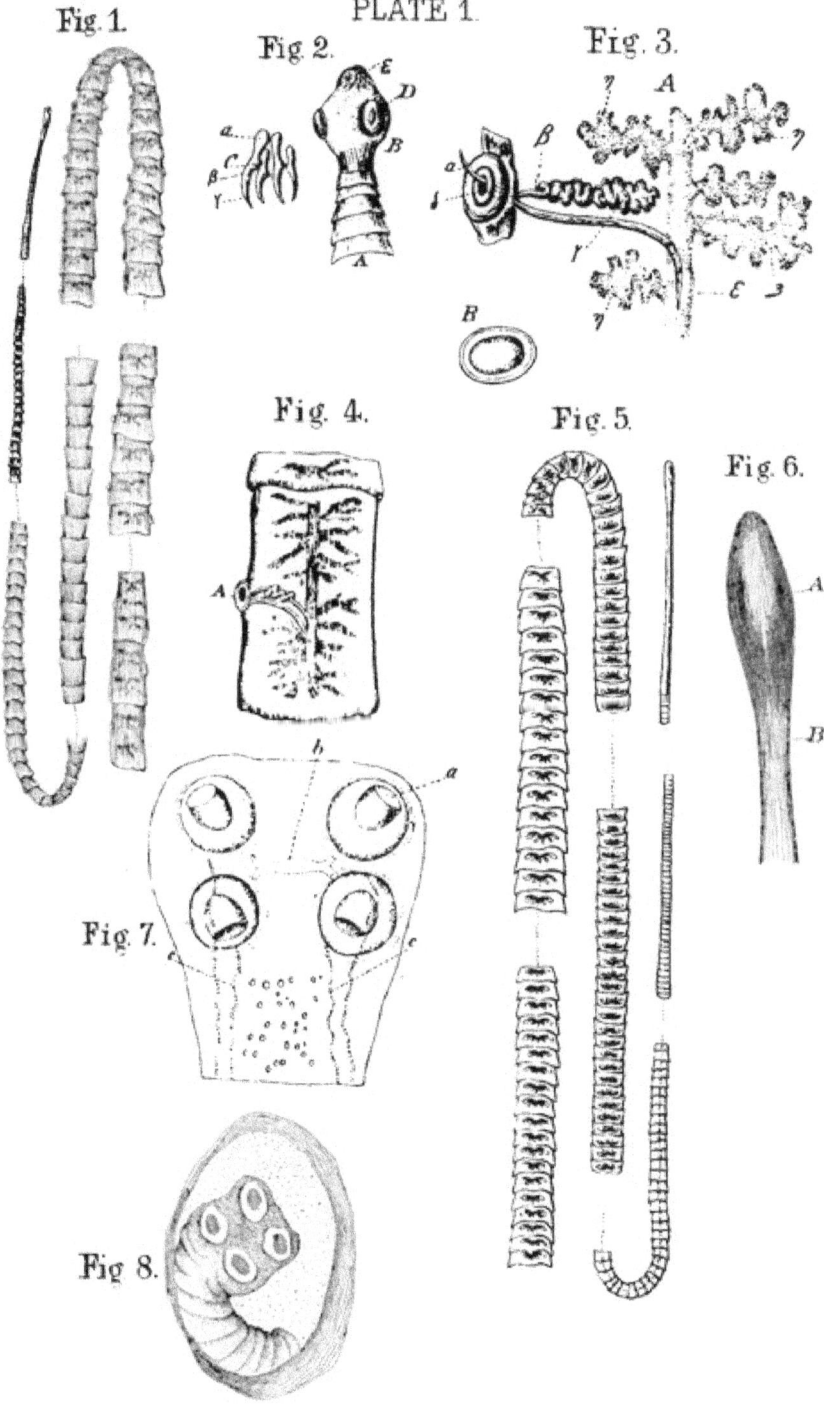

PLATE 2.

Fig. 2.

Fig. 1.

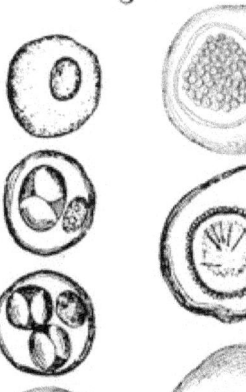

Fig. 3.

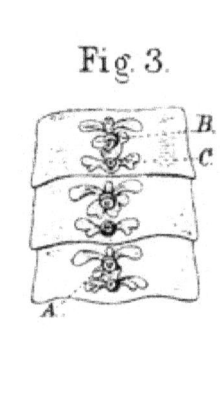

Fig. 4.

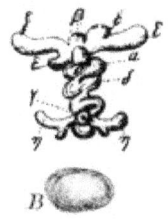

Fig. 5.

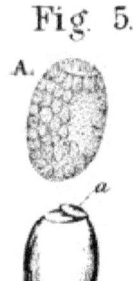

Fig. 6.

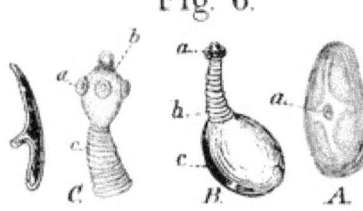

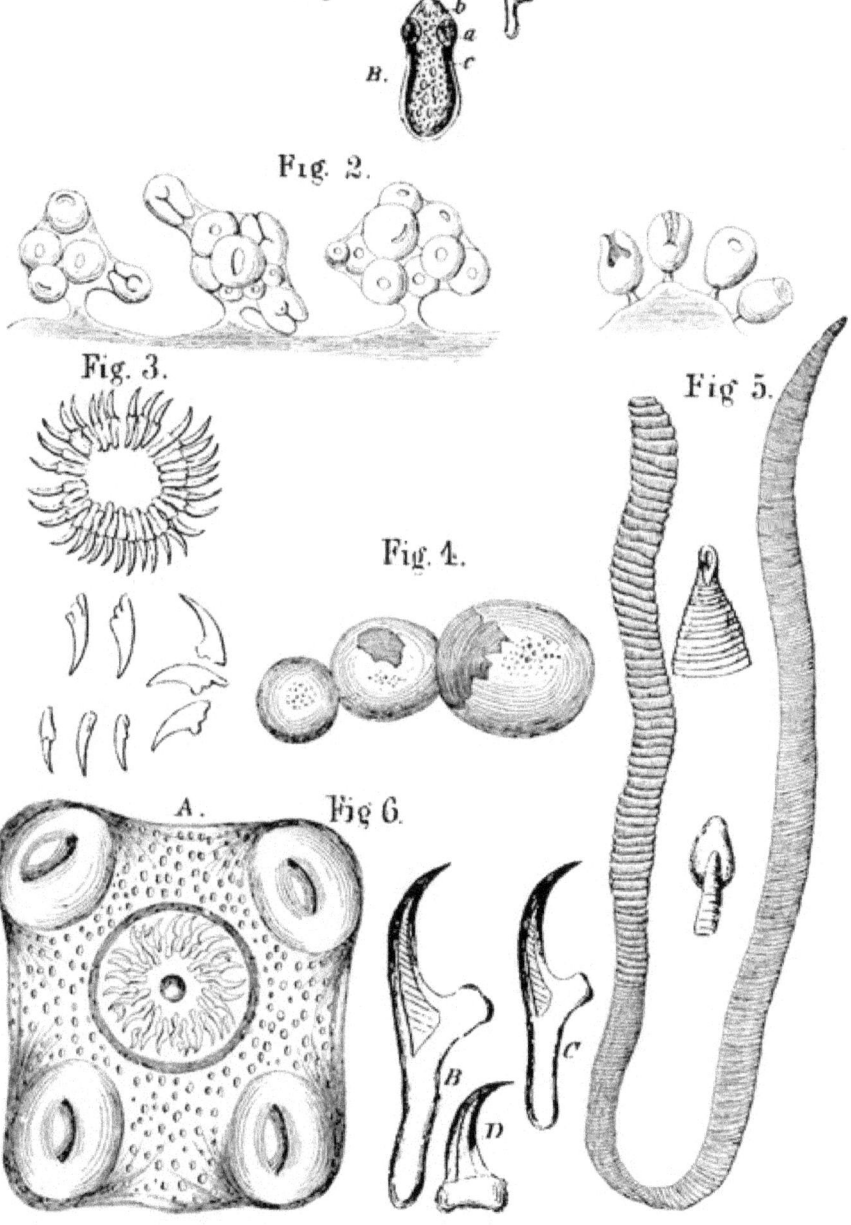

www.ingramcontent.com/pod-product-compliance
Lightning Source LLC
Chambersburg PA
CBHW031605110426
42742CB00037B/1222